YouTube

How to create great content, grow a following, and make money on YouTube

Jacob Kirby

Table of Contents

Introduction

Makeup tutorials, gaming channels, sitcom reviewers, language teachers, medical advisors, ASMR, musicians, social awareness experts, comedians, and much, much more — all this range of content can be easily found on YouTube, the most popular video platform of the past couple decades. With so much variety and diversity of topics, many people feel encouraged to produce their own video content and maybe even build their own subscriber base. If you are reading this, you are probably one of these people (or at least you are curious enough to learn how the whole thing works!). But would this even be possible? Could you also become an influencer? What would it take? Where to start?

This book answers these and other questions and helps you begin, improve, or monetize your channel in the best possible way. But don't get too excited, because success doesn't come overnight! It takes dedication, study, and investment to get there. For this reason, we will divide this manuscript into three major parts.

First, we will start from the beginning, with tips on how to start a channel, how to get comfortable talking to a camera, and how to choose the best equipment to create your videos. Once this is covered and the content is settled, we will move on to the second part, which concerns tips for growing your channel, gaining followers, and understanding how the so-called "YouTube algorithm" works. This will help you maximize your creative potential and attract more and more people interested in what you have to say. The third part will be more business-oriented; who hasn't imagined themselves making money online? Setting up your working hours, not having to commute to the office, talking about whatever you want and to people who also like the same things as you and who are interested in what you have to say — it certainly sounds like a dream! Making money from videos is a potential reality on YouTube, but you have to be smart about it. The last part of this book will delve a little deeper into the details of video monetization and also partnerships with brands and companies that want to advertise their products. If you do it the right way, you will have the possibility to grow on the platform without harming your image, as maintaining your authenticity and honesty with your audience is highly important!

Taking the first step is often the hardest part. Many people are afraid, as they don't think they are communicative enough or are camera-shy, but you will never know what it will be like if you never try. That's why it's important to act as soon as possible. Don't worry about having the best camera, the best lighting, or the best setting; if you gather enough courage to put yourself out there and let people know who you are and what you have to say, the rest will come with time.

By the end of this book, you will have all the information you need to create authentic, informative, interesting, and ambitious content for YouTube. You will know exactly where to start and how to upgrade your resources as more people hit that subscribe button. This book intends to serve as a guide to be consulted as you reach new milestones in your YouTube journey and have to deal with different scenarios. So, let's start from the beginning!

Chapter 1: Getting Started on YouTube

So, you want to create a YouTube channel. You are full of ideas, hopes and dreams, but it is difficult to know where to start, how to organize yourself, and how to prepare something that people will watch (preferably for several minutes straight). This introductory chapter is dedicated to beginners who are still unsure about where to start and are looking for practical tips on how to set up their video publishing business. In the next few sections, you will learn exactly how to choose a topic, plan your content, and style your channel in the best possible way..

Choose a Niche

With so many topics on the rise, there's no doubt that deciding on the perfect content for your digital platform is difficult. You may think that everything has been done before, from daily vlogs to viral pranks. How do you even choose a subject or topic to focus on?

Many people start by talking about their passions and adapt their content as they grow. A perfect example of this is veteran YouTuber Anthony Padilla. When he started in the late 2000s, he was just some guy from Vine who enjoyed making jokes, but who later started opening up a little more about his life and personal struggles. After ten years on the platform, he began a series of videos titled, "I spent a day with..." comprising interviews with people from the most diverse personal backgrounds (ranging from LGBT activists to people who identified as witches). This proved to catch people's attention, and today, that is the content he is mostly associated with. Within a decade, Padilla moved from silly teenage jokes to more mature content that kept hold of his audience's interest.

The bottom line is this: no matter what you choose to create, don't be afraid of experimenting, changing, and adapting your content. You are not required to produce anything that doesn't suit you or that no longer works for you. To find your (first) niche, try thinking of something that you enjoy very much at the moment. The content to be created should come from a passion or something that drives you to speak with confidence. If you must, make a list of topics that motivate you to talk for hours straight.

Suppose you are obsessed with stories of all kinds. You want to talk about all of them, your theories, the books they were based on, fun facts, etc. Perhaps this is too broad a subject. Fans of sitcoms from the 90s are not always the same ones who like to read romance novels. So, it is worth considering "filtering" some of your content — for example, limiting it to fantasy stories (books, movies, or series) or types of media (e.g., visual content, like TV shows and movies). This helps you maintain a subscriber base that will stay with you longer and watch most of your videos.

If you opt to talk about many subjects, it is important to create something that is an integral part of all your videos, like a sort of "signature" or "trademark." For example, in the channel "Amanda the Jedi," Amanda talks both about romantic teenage novels and sub-celebrities who expose their children on the Internet or get carried away by money. Now, if you stop to think about it, those pieces of content seem to be very different. How does Amanda keep her subscribers interested? By adding her personality to every video. Many of her subscribers are not even there for the content itself, but her opinions about such content. The way she approaches her topics is full of humor, irony, and reflection. This entertains people because they probably think the same, but cannot express themselves with as much spontaneity and sharpness as she does.

This brings us to the second step of creating your channel: understanding what makes you stand out. Now, there are thousands of channels on fantasy stories. It is something that has been done before. So is the case with make-up channels. And ASMR channels. And reaction channels. And gaming channels. And anything else you can think of. That's why it's so important to be authentic. Your personality will be the essence of your channel. This is what will differentiate your channel from the thousands of others like it.

But don't panic! This does not mean that you need to be a genius of humor for your channel to succeed. Many things can be good ways of setting you apart from others — even if they are more subtle. For example, makeup artist Nikki Tutorials separates herself from her peers for her makeup style, but also by including a "Dutch word of the day" at the end of each video. This way, she can share a bit of her culture with her subscribers, and it has become a distinctive feature of hers.

Besides the "Dutch word of the day," Nikki also opens up about her insecurities while putting on her makeup. She has already discussed her difficult relationship with her body image and self-acceptance, and she always enjoys reinforcing that make-up is a powerful ally in her journey. This makes the audience feel closer to her and see her as a friend. By alternating personal vlogs with the major topic of the channel, a make-up channel like Nikki's, for example, stops being "more of the same," and instead becomes about the person who is behind the camera. This helps break with the idea that makeup should hide something, and it celebrates it as art. People feel connected to personal videos; it conveys charisma to the viewer. And charisma alone is a powerful differentiator.

If you are not sure how to add something different to your channel, you can do a little reflective exercise and ask your friends and family what they think are your strongest points. You may be surprised when you find out exactly what makes you stand out in their eyes!

For instance, suppose you asked 3 friends what they think your strengths are and you end up with the following answers:

1. Intense in all your passions (because when you talk about what you like, you don't contain your excitement).

2. Empathetic and care about other people's feelings.

3. Love helping others.

With this information, you can narrow down your options to a few possibilities. You can choose one of your passions as the main topic, and:

- Teach people about this topic from your perspective and experience.

- Help people understand the topic better (even if it is general, like "pop culture" — you can create theories, explain plots, break down characters, etc.).

- Tutor people on the topic.

- Comment on the topic considering the perspective of someone who is just getting started, etc.

Plan Your Content

Once you've chosen a niche to focus on, it's time to plan your content. Consider researching trends and recurring topics of the moment. It doesn't matter the nature of your channel; being aware of updates and knowing what is going on in the world is always a good idea to come up with fresh insights. A good example of this is Lorelay Fox, a Brazilian drag queen who used to turn her YouTube guests into drag, but who, with the COVID-19 pandemic, needed to seek new alternatives. She started researching the trendiest issues of the moment and, based on whoever people were talking about the most in a week, she digitally transformed them into drag. With only an iPad, a digital pen, and the Procreate app, she has already turned reality TV stars, former presidents, and fellow YouTubers into glorious drag stars, and the list is only growing.

It is important to see what's trending, because the less time it takes for your video to be seen, the more the algorithms will benefit you. Subjects that are being discussed by everyone and on all social media channels are popular and attract more attention from the public. You can use this to your advantage. Keep an eye out for new products that can be related to your channel, and don't forget to check social media regularly. Instagram and Twitter can be significant sources of creativity, so follow accounts that keep you up to date with everything that's going on around the world. Check out the trending topics on

Twitter — everything that happens around us can serve as a reference for a video (yes, even the most random things). For example, if there is a new show that everyone is talking about, is there a way that you can connect it to your channel? By blending two unrelated topics (your niche + popular TV show), you are likely to create something unique to you and your channel.

There's no room for competition on YouTube either; following other content creators who are in a similar niche as yours is also a great inspirational source — as long as it remains inspirational and doesn't turn into plain plagiarism. Being a YouTuber is like having any other job. Connecting with your peers and engaging in some networking can only strengthen your content and help you grow (that is, as long as you make the right choices — stick to your values!). Be sure to consider collaborating with other channels, even if you're both still small. This will give more visibility both to you and to another creator, as well as broaden the range of content for your channel.

Style Your Videos

Just as important as the content you create is how you decide to present it. So, make sure you take the time to schedule a regular frequency of uploads, think of interesting titles, and style thumbnails for your videos. You may think that these are small and unimportant decisions, but they can help you gain more views, and this is helpful at the beginning of your channel because it makes people curious enough to click on your video and learn what you offer. It used to be enough to post one video per month with a generic title and a frame taken from the video itself, but that time is gone.

People are attracted to things that are aesthetically appealing, and video promotion is becoming more and more visual. There is a reason people are on YouTube — they are looking for something that will catch their eye, not something boring or visually cluttered. You don't need to be a design genius or an expert video maker, but putting some effort into these details can change the perception of your channel drastically.

Try to start by posting at least one video per week. If you have time and are bold enough, consider posting two. This frequency helps keep your channel in the recommended videos tab and keeps your subscribers interested in your content. If you post a video once and then suddenly disappear for three weeks, chances are that people will forget about you — and so will the YouTube algorithm. Set up a schedule to organize the videos that you want to post in a month, and consistently upload them on the same days of the week — preferably at the same time. Posting the video at a peak hour, when more people might watch (when they are more likely to be home or free to browse the internet, for example), also helps your video rank higher in views.

Your title doesn't need to be extremely sensationalist either, but it should draw the viewer's interest. Some popular title choices involve a short first sentence of one to four words, followed by a colon and a brief explanation. See the examples below, taken from actual videos:

Dog and Cat: Obstacle Challenge (MilkyBokiTan)

Blind vs. Sighted: Who has better senses? (Molly Burke)

Another popular approach is to ask a simple question or make a kind of promise in the title, as this instigates curiosity and makes the viewer want to find out the answer:

Why does Pepsi have a game show? (Drew Gooden)

How to do a GERMAN ACCENT? (DontTrustTheRabbit)

If you struggle with anxiety, this mind trick will change your life (Mel Robbins)

This is what an astrophysics exam looks like at MIT (Tibees)

Simple, self-explanatory titles also work:

How I Wrap Gifts as a Blind Woman (Molly Burke)

The YouTuber who faked a mental illness (Primnik)

And videos in list format convey the idea of something fast and that the viewer can follow more easily:

15 American things that Europeans find weird (Get Germanized)

5 common habits that make people instantly dislike you (Charisma on Command)

Finally, there is also the option of putting your opinion on a certain subject in the title. This makes people with similar opinions tempted to watch and see whether they agree with what you are saying:

Descendants doesn't make any sense... (Alex Meyers)

"365 Days" is WORSE than 50 Shades of Grey - Explained (Amanda the Jedi)

The thumbnail of your video is a great ally for your title. Many YouTubers use the thumbnail as a sort of subtitle, putting a "catch phrase" on the image selected to be the cover of the video. This avoids extremely long titles and still allows you to increase the chances of catching the viewer's attention. This combo can make it more likely for a potential viewer to click on your video over someone else's.

YouTuber Cydnee Black, for instance, has a video titled "Why My Eyes are Blue, Changing Your Eye Color." Now, judging by this title alone, some people would simply think that this is something normal or a case of contact lenses and move on. But Cydnee used the thumbnail to her advantage and, besides adding a picture of herself, added the phrase "Black Girl With Blue Eyes" to it. This combination is much more interesting than the title alone, as it instigates the viewer's curiosity and suggests a detailed explanation regarding the video.

Chapter Summary

In this chapter, you have learned:

- How to take the first steps to start a YouTube channel.
- Tips to choose a topic for your videos.
- Ways of planning your content.
- The best strategies to come up with appealing titles, thumbnails, and style for your videos.

But this is only the beginning! In the next chapter, you'll learn a little more about the process for creating the perfect lighting, choosing a camera, and how to get comfortable shooting your video. Read on to find out how!

Chapter 2: Lights, Camera, Action!

The lighting and recording of your video are just as important as what you have to say. They can be the determining factor in whether or not the viewer will stay until you are done talking. Nowadays, appearance matters more and more, and technology is progressing fast, so a video shot in dark surroundings or with low audio is unappealing to most people. In this chapter, you will learn more about the technical parts of video creation. After every section, you will get closer and closer to becoming your own director of photography, producer, and cameraman!

Prepare the Equipment

All right, the word "equipment" can sound pretty intimidating. But we are not talking about spending piles of money on typical Hollywood super-production equipment here; it's all about working with what you already have or what you can afford, and making adaptations accordingly. For example, if the only camera you have is your phone, it's fine to use it to record your videos, but you must always be aware of potential obstacles and take action to remedy them.

No matter which camera you choose, you must keep only one thing in mind: your image may look bad, the lighting may not work, the scenery may be horrible, but poor audio is the most unforgivable. People may watch a video with a 140-pixel image, but they will immediately close the window if there are significant flaws in the audio. So, if you don't have a good camera, just work on making sure the sound is clear in tone and comprehensible. No need to study advanced settings on your video editor either; you can do this by simply connecting a headphone to your device. When you do this, the sound comes from the built-in microphone. The trick is to keep it as close as possible to you and to speak in a clear voice. You'll be easy to hear, and people will be far more likely to keep watching despite any other imperfections in your video settings.

This leaves us with a "basic kit" of just two items:

1. Phone or camera.
2. Headphones with built-in microphone.

And voilà! You are ready to go! Unless, of course, you want to fix that lighting...

Set up the Lights

If you are unhappy with the quality of your video, even if the audio is within acceptable comprehension levels, a good way to improve it is to alter the lighting of your shooting environment. While it's true that bigger YouTubers normally have a ring of light and soft boxes, if you don't have access to this kind of equipment, you can still improvise with cheaper options. Good lighting can be created with sunlight or inexpensive lamps. Natural light can sometimes be just enough if you situate yourself correctly.

If you opt to shoot with no equipment, choosing a time of day when there is more light in the room can give you amazing lighting, but you need to pay attention to a few details. For example, you must never shoot with the camera pointing towards the light; instead, be the one facing the window, so that your body is completely illuminated, and the camera can easily focus on you.

Don't worry if it's not sunny — the important thing is to get light into the room, which is possible even in cloudy weather. Of course, this method is prone to light inconsistencies from video to video, but this shouldn't be something that puts off your subscribers. It's a valid way to improve the quality of your image even if you have a camera that is not as good as you would hope.

If you prefer to use artificial light to maintain a similar lighting pattern in all your videos (and also to not be restricted to filming only during daytime), there are inexpensive alternatives to professional lighting equipment. Something that works well is to choose bulbs that don't have such a strong brightness. Yes, you read that right: instead of buying one large, expensive, high-intensity lamp, buy a set of six or eight cheaper, weaker lamps, and add bulbs that are not too strong. Then arrange the lamps in a circular or square array as best as you can.

For channels that require your face to always be in the spotlight (e.g., makeup channels), medium lights are most effective. If possible, arrange them like the ones you see in the dressing rooms of big TV stars: in a square pattern, framing your face. You also don't need to buy lamps individually, as you can instead opt for string lights.

Of course, bright lights are not out of the picture. If you prefer stronger lamps, you can buy a single one and cover it with wax paper to soften its intensity a bit (you don't want your face to be so bright that it becomes unrecognizable).

Believe it or not, some YouTubers improvise so much in lighting that they even create DIY light reflectors. So don't be discouraged if your own lighting isn't perfect right away; start with what you have, and as you grow you can invest more in the equipment.

Talk to the Camera

Talking to a camera can seem strange at first. After all, shooting a YouTube video is nothing more than talking to yourself for a few minutes. Most people are not used to speaking out loud unless they are sure that there is a counterpart to respond to and establish a conversation with. For this reason, you may feel a little inhibited when facing the black lens of the camera. What if the neighbors hear you? What if your roommate thinks you are talking to yourself? What if, what if, what if...?

The first step to speaking well in front of the camera is to not let yourself get intimidated by it or by how weird it feels to talk when no one is in the same room to hear you. The lack of communication is just an illusion; once your video is published, you will be speaking to anyone who wants to hear you. So, when you look into the camera, try to forget that you are talking to a lens, and imagine that the black circle is an eye, a pupil that represents hundreds of people. This is especially important because it forces you to look into the camera itself, and not at your image, the picture on the wall, or other distractions in the room. Fixing your gaze on the lens will make whoever watches your video later feel as if you are looking at them and talking to them in a normal conversation.

To make your video feel even more like a conversation, you can also adopt some techniques that will help you be less stiff in front of the camera and loosen up a bit more. One thing you can do to act more naturally is to simply pretend that you are in the company of close friends. Talk to the camera in the same way you would talk to them, with all the honesty, spontaneity, and jokes that you would normally make. This is not so hard to imagine when you think about it, as technically, whenever we make a video call, we are also talking to a camera! The only difference is that the person we are talking to is also speaking to their camera whilst listening to us. It's all a matter of perspective, and over time you will learn to look at the video recording process more naturally.

If you feel it is too difficult to shoot and that, even after trying to speak in a louder voice, you are still inhibited and apprehensive to act as you normally would, you can try shooting a more relaxed video, free of expectations: no script, no topic, just you talking to the camera about whatever comes to your mind just to get used it. This takes the burden of "perfection" off your back, and you'll be free to be spontaneous and natural.

Remember to be yourself. Creating a YouTube "persona" rarely works. Being the person that you are off-camera simplifies your creative process and allows you to show yourself exactly as you are in all areas of life — whether on or off social media. People like authenticity because being genuine allows others to see themselves in you, share similar stories, and have a sincere and valuable exchange with you. Creating vlogs, for example, helps to strengthen the authenticity and identification between viewer and creator.

No matter what your profile is, introspective or spontaneous, shy or outspoken, it is possible to be a brilliant content creator. Just be patient and start with baby steps; you will grow little by little, so don't get your hopes up too high right away. Everything is a process, and every process takes time. Those who want to start off enormous right at the beginning are doomed to give up before they even grow.

Chapter Summary

In this chapter, we have covered some strategies for improving the quality and appeal of your videos, as well as your channel. These strategies include, among other things:

- Ensuring quality sound for your video, even if the image is not of the highest definition.
- Preparing a starter pack, which includes a camera and headphones to shoot your videos.
- Strategies to improve the lighting of the room, including both natural and artificial options.
- How to feel more confident and speak more naturally in front of the camera.

These first two chapters complete the introductory part of the video creation process. Over the next few pages, the focus will shift to ways of growing your channel, and how you can profit from it.

Chapter 3: How to Grow Your Channel and Get More Subscribers

When you know what kind of content you want to produce, how often you want to produce it, and what kind of equipment you need to do it well, you can start thinking of ways to maximize the potential of your channel and attract the attention of people who may relate to what you have to say. The following tips will help you achieve just that; you should read them carefully and then choose the ones that best fit your content and your availability.

Produce for Multiple Platforms

If you want to create content on YouTube, there are usually reasons why. It is a platform that enables you to solely focus on videos, whether you create scripts or prefer more freedom in the creation process. But being active on other social networks can also increase your visibility and spark the interest of people who might not otherwise find you on YouTube.

This does not mean that you have to devote time to creating new content for each of these platforms (after all, that would take up many hours of your day), but that you can use them to promote your videos. You can also create content only for

YouTube and then adapt it for other platforms. For example, use Instagram stories to announce a new video, or post a photo on your feed related to your channel's content. Take an interesting clip of your video and post it on other social media to get some attention. Add hashtags related to your content so that the right people can find you.

You don't have to mix your personal profile with your digital content profile, either; you can create professional accounts to promote your work. Just keep in mind that different websites and apps operate differently. On Instagram, for example, you must be very active so that you can keep in touch with your followers and continue to show up in their feed. If you disappear for too long, you likely won't reach all the people who crave your content. For you to keep appearing to everyone, your posts need to have proper engagement. On Instagram, numbers are more important than almost anything else. Poorly engaging photos or stories are likely to be at the bottom of the "algorithmic priorities" of people who don't actively search for your content. An excellent technique to maintain engagement is to create polls or questions in your stories. This encourages interaction, keeps people interested, and also allows you to learn more about them, why they are watching you, and what kind of lives they lead.

On TikTok, the process is a bit more challenging. Although it is also a video platform, the people there are looking for something fast, entertaining, and less interactive. If you want to establish your presence on this network so that more people will find you on YouTube, you may need to appeal a little more to creativity. TikTok's content is usually informative, straight to the point, or focuses on entertaining. If you have a slightly more "serious" or inhibited profile, perhaps this network is not for you. But if you create short promotional videos, keep in mind that clips that involve role-playing and make-believe are popular on this app. It's almost like a theatrical stage where you can explore your acting skills. For example, it is common to post "this or that" videos (where you place two options on opposite sides of the screen and move left or right to state an opinion) or made-up dialogues. This can be a fun way to adapt your content and attract more people to your channel.

You can also use Twitter to get inspired and share your videos. However, Twitter is much more about *you* than about your work. Posting a series of repetitive tweets with links is of little interest to users. Use this platform to express opinions, to help people with informative threads about your content, and also to share your videos. This way, people will identify with you and will be more likely to check you out on other social networks.

On the internet, people can be as direct as they want, often inspiring cruelty, and we know how that can damage people's mental health. If you feel you are more sensitive to this kind of interaction, don't force yourself to stay on these networks. Know your limits, and keep in mind that if being on so many apps is making you anxious or causing you to self-sabotage, it's better not to be on them at all. Your health and well-being should always come first!

You must also remember to be professional and transparent whenever you post your content on other networks. Platforms like Instagram provide much greater proximity to followers and allow for more extensive interactions. For this reason, you may also be more exposed there. Therefore, it is so important to show honesty, without appealing to clickbait or advertising something in the wrong way. Show your content the way it is, and those who become interested will join your channel to stay.

Because content creators are their own bosses and take care of every part of the video creation process, many people feel as if they have control over their "fans." But the reality is just the opposite; it is your followers who make you grow professionally. If you don't give your audience what they want, they will simply abandon you at some point, and you will need to change your methods much more often. Therefore, it is so important to establish principles and stick to them. Those who agree and identify with you will be more likely to be loyal followers of your work.

Collaborate with Other Influencers

When establishing yourself on other apps, follow people both outside and inside your niche. Outside because it helps you "get out of your own bubble" and stay on top of what's trending on social media. Inside because then you have chances to bond with other creators and get inspired by them (*get inspired* to make your own content, not inspired to steal theirs,) and even collaborate with them. Your YouTube peers are not your enemies, but your allies! A community of creators can help each other and also learn from each other. Therefore, it is very important to find people who are in line with your principles. These colleagues can share their experiences and produce content with you.

There are different collaborative works on YouTube, and each one can have a different impact on your subscriber base. No matter the size of your channel, collaborating with both large and small channels can be helpful for you.

If your channel is small and you collaborate with another small creator, you both increase the chances of gaining more subscribers. In that case, one helps the other grow, so it doesn't matter if your channel is about Norwegian culture and the other channel talks about games. When the other contributor is as big as you are, it may even interest you to select someone who does not have a niche as similar to yours, as there will be more chances to reach people who would not find you otherwise. Regardless of the channels being similar or not, the chances of growth will be similar, and you will have little to lose. Just be careful not to accept just any offer; it is worth considering if you actually like the person's content and if you think your subscribers would also be interested in it. Always have this as a principle: if you were subscribed to your channel, would you want to see this? If the answer is "yes," then it is something worth considering.

Collaborating with bigger channels is always a great opportunity, but it can also tie your public image to that influencer, so it is worth looking at these possibilities more carefully. You don't want to collaborate with a large channel only to find out three days later that they have been criticized for making exploitative, cash-driven videos. Explore the person's channel closely and think about whether you would like to have your image attached to that person. Ask yourself if you resonate with their content, and whether you would actually like to meet them, or if you would be only doing it out of impulse. You can always refuse offers.

If you are the person who has the larger channel, you may wonder what you might gain by collaborating with someone small. There are many reasons large creators join forces with smaller channels, and they usually come down to the fact that they agree with your content, are interested in it, or believe that you have the potential to reach more people. This is also a great exchange experience where the "big" channel can learn from the "small" channel and vice-versa.

No matter the nature of the collaboration, the most important thing is to know your reasons for choosing this partnership. Remember that it's not just about gaining more followers, but also a great way to express your creativity, spread information to different audiences, share important causes, or help another creator grow. Yes, you'll be gaining more engagement and yes, it's almost a kind of additional marketing, but it's also about making friends, having fun, and varying your content a little.

Understand the Numbers

Paying attention to the numbers can also help a channel grow. On YouTube, you have access to all the statistics for each video. Look at each one and check which of them have been the most successful so far. If there is a style of video that receives significantly more views, this means that you have made something that worked, and that people liked it enough to share it with others. It is also worth observing if you suddenly gain several subscribers after posting a particular video. Every time someone subscribes to your channel it is because they liked something, and if this happens shortly after you post new content, you can bet that people expect more of that type of work from you. If you post a video of interesting facts about your job, for example, and this is your most successful video, posting something about your morning run may not garner the same number of views.

Besides the number of views, you can also look at the statistics to see what parts of the video your audience was paying particular attention to. YouTube allows you to see exactly what your subscribers like and dislike — it's up to you to learn how to interpret this data.

Learn to understand the numbers. This helps you figure out exactly what your audience wants. They are the ones driving you forward, so listen to them. It's not uncommon for the next viral video to come from a suggestion in the comments. Your viewers know what they want to see, and taking their opinions into account is good for both you and them. If you deliver something your audience likes, they will reward you with more engagement.

Chapter Summary

This chapter has covered a few strategies to grow your channel and reach more people. It is important to remember that the tips offered here are not your only options, and there are always alternative ways to attract subscribers to your channel —

especially with the rise of new social platforms. Here, we focused on some key actions, namely:

- Producing for multiple platforms, adapting content for other social media, and promoting your videos on the apps that suit you.
- Following other influencers with similar or distant content to yours for inspiration, networking, and collaboration.
- Collaborating with these creators to increase engagement on both channels and to reach new audiences.
- Understanding the significance of numbers and using them to your advantage.

Similar to this last topic, in the next few pages, we will delve a little deeper into the world of algorithms in order to understand how YouTube recommends your content, who it recommends it to, and what makes it appear on the main pages of the platform.

Chapter 4: How to Rank in the YouTube Algorithm

As we have already seen in the previous chapter, understanding the algorithms of the social network you are part of is essential if you want to attract more people to see what you offer. In this chapter, we will list some of the most important steps for ranking in the YouTube algorithm. There are certain details that, if ignored, can significantly reduce your reach on YouTube.

The algorithm allows you to be seen. It serves to create a basis for analysis, patterns, and behaviors among different clusters all across the platform. For every user who visits YouTube, certain types of content are filtered for them based on the channels they watch, their preferred video lengths, and how they interact with each type of video.

There are a few strategies that YouTube uses to determine the algorithms and recommend the right videos for each viewer. The major goal is to find the content that best appeals to every user and to keep these users entertained as much as possible. To do this, YouTube thoroughly analyzes user activity and creates personalized pages according to its profiles of users, with search results and recommendations differing from person to person. The title of your video, the description, and the keywords are crucial for you to appear in a user's search since the words chosen must match the profile of the person who will find your video. The number of likes, comments, and views also contribute to your video appearing at the top of the search results.

Based on your subscriptions, notifications, trending videos, homepage, and watch history, YouTube schedules the best content for you to receive. Knowing this, you can and should make it easier for viewers to find you.

Analyze How People Interact with Your Videos

When people say that all kinds of interactions with your videos are important, that's no understatement; the number of likes, dislikes, subscribers, comments, and shares all help you gain visibility, because that's how YouTube understands that your video is relevant. No wonder practically all YouTubers drop that famous phrase urging viewers to like their videos, hit the subscribe button, and activate notifications to know when a new video is published. Even negative interactions are a way to promote your video. Angry people are just as likely to share content they hate as people who share content they love.

Also, not only do these interactions say a lot about your video, but they can also tell you a great deal about who is watching them. YouTube Studio has an "Analytics" tab. Here, as well as in the details of each upload, you can check what were the highlights of each video, which parts were most watched, how long people stayed watching the video, etc. This data is excellent for assessing what type of content attracts the most attention from your viewers and helps you better understand what works well (both for you and for them), and what would be best to change. It's a way of getting to know your subscribers better and delivering content that is more aligned with what people want to see.

But how do you know if a video is doing badly? If you only see a lot of numbers and are unsure how to make use of them, try comparing the ones in newer videos with those in older ones. For example, if you usually have 100 views, and this time you only have 30 views, it is a sign that you need to make some kind of change for your video to reach more people. When this happens, pay close attention to the details of your video and see what can be improved. For example, maybe your title is not catchy enough or your thumbnail is uninviting. It is also a good idea to set aside a few title and thumbnail options even before uploading your video; you can create 3 backup titles and thumbnails, for example, and then experiment with them to see which one performs best.

Another thing you can do when engagement is low or there are no comments on your video is to be the first to comment yourself. You might think that commenting on your own post is practically the first sin of the social media testament, and while this may be true for that selfie you posted on Instagram or the like your grandmother left on her own photo, this is not as disgraceful to do on YouTube. When people like you, they expect to see you commenting. So don't be afraid to interact with your video! Include a question in the comment section to encourage a discussion among the comment lurkers.

Take Hints from the Comment Section

Speaking of the comment section, it can tell you a lot about why your video is not doing well or why it is in a worse position than most of your other uploads. Sometimes, even changing the thumbnail or the title doesn't make any difference. So, it is also important to read the comments. While it's true that many content creators are sometimes anxious about what others are saying about their video (after all, YouTube commenters are like a boss giving negative feedback, only on the internet it all escalates quickly and it's like they are yelling at you instead of behaving professionally), constructive criticism can be very helpful.

But be aware that we are not talking about haters here. A hater's job is just to trash your work without giving you any tools to defend yourself or improve your work, while actual criticism can help you grow. When someone says something negative about your work while offering you alternatives for improvement, you can take this as something positive and consider what they are saying. Often someone shows up in the comments to give you a hint of something that can be improved. For example, if someone writes, "I don't understand this title," or, "Your audio is low," or, "Speak louder," you clearly know what the issues are and what you should change.

These tips from your viewers don't just extend to YouTube either. Besides checking your video metrics, you can also track how people are interacting with your posts on Instagram. On this platform, you can see how many people have saved your post or forwarded it to others. These numbers help you understand if your content is interesting enough to be shared with more people, and you get to know what is going right or wrong with your channel.

Optimize Your Video Description

As a YouTube user yourself, you might not even look twice at the description of a video, since you're more interested in watching it or reading the comments instead. But in reality, this space can be very useful for promoting your video. This is where you will need to put your marketing knowledge to use and add some clear and concise sentences talking about your video. Always add some keywords so that your chances of being found in the search engines increase.

You don't have to be an SEO genius, but it also doesn't hurt to do a little research on some tools that can help you maximize the reach of your video. Some websites such as answerthepublic and keywordtool can help you look up the most popular search engine words relating to your video and can be real allies.

Thinking about keywords may seem easy, but when you need to fit specific words into a sentence, it is very common to lose the natural feel of the text; make sure you create sentences that have a good flow and allow the reader to get all the necessary information. Ask someone to read it for you and get a second opinion if you can. You can also choose one or two keywords and repeat them in the title to double your chances of being found.

Remember to also create a description for your channel. Choose the keywords that you think have the most to do with your content, and make sure that these words are in the first few sentences of your description. YouTube's algorithm prioritizes the first few sentences because they are what encourage people to keep reading and click on "view more" when they find your channel or video in search results.

Never forget to add value to your descriptions. If they were just for adding keywords, you would only use hashtags and wouldn't even have to worry about writing a description. Your descriptions should be practical and address the needs of the viewer. For example, if you are going to teach how to prepare a recipe in your video, your description should contain a summary of the content so that before people even click on the video, they know what to expect and feel compelled to follow your recipe. If you simply fill your description with keywords that are not embedded in an easy-to-read sentence, your description becomes much less attractive, and people might find other similar videos that seem to be more appealing regarding what they are looking for. Imagine reading a description that says "tomato pie cheese dough recipe cooking cook vegetarian." Would you find this appealing in any way? Most people would not.

This is not to say that you should not use hashtags at all; in fact, they can help to promote your video, but avoid overusing them. YouTube is becoming increasingly good at detecting irrelevant content, and the more hashtags you use, the more likely the platform is to take that as misleading content. A good rule is to stick to a maximum of 15 hashtags.

Pay Attention to the Length of Your Video

Years ago, when YouTube first started, the tendency was to upload quick videos up to 5 minutes long. Anything longer than that usually made people lazy, and they would close the video without finishing it out of boredom. People joined YouTube to watch shorter content, without spending many hours on the platform. But with the rise of digital influencers and other short video apps, this trend has changed drastically.

Compared to the tiny clips of just a few seconds long on TikTok or Snapchat, even the 5 minute "short videos" on YouTube have become comparatively long, and more and more people have looked for longer videos on the platform.

Today, YouTube is a "long video platform." People can spend hours and hours watching videos that are 30 minutes, 40 minutes, and even over 1 hour long. If you pay attention to it, you'll notice that it's quite rare to come across anything shorter than 12 minutes on YouTube nowadays. The fact is that for YouTube, the longer you can keep the person watching your video, the easier it is to convince them to subscribe to your channel and consume more of your content.

Still, length is nothing without excellent quality content. In the early years of YouTube, it was also much more important to get a lot of clicks rather than to keep users engaged. Today we know that keeping people interested means more than the number of views, even if your video is not that long. This change has occurred because of the extensive amount of clickbaiting, which is what happens when the video suggests something completely misleading and non-existent in its actual content, just for the sake of getting more clicks. YouTube has also made it harder to post short videos that are only uploaded to keep regular content, as well as videos that are only long because it allows for more viewing time.

So, here's the lesson: don't try to mess with the YouTube algorithm. Learn to produce relevant content that adds something to the users' experience and that they will watch. With the number of views losing its relevance, the algorithm requires you to keep your viewers entertained for the entire duration of your video, instead of turning it into something long (but isn't watched) because it's not interesting to viewers.

No matter whether you prefer to create shorter videos or long ones, more important than that is to make sure that the user experience is valued. Here are some creative tips for both short and long videos:

Short Videos

With short videos, the best way to start is by hooking your audience with a question or something that will have been explored by the end of your video. Remember, the time watched is more important than the number of views, so if you spend 2 minutes of your 8-minute-long video introducing yourself or covering something that is not related to the main subject of the video, people might get discouraged and not feel like what you are saying is worth their time. So, keep introductions short and try encouraging people to keep watching by making them curious about what you have to say.

Long Videos

If you want to create longer videos, a tip that can work well is to build analogies. For example, if you have an educational channel, this can bring your audience closer to a subject that is considered more difficult, or somewhat academic, by connecting it to something your audience already knows about or that is part of their world view. People rarely seek educational videos on YouTube that look similar to a boring university class. By adding something different that feels more like a conversation than preparation for the SATs, you are more likely to grab your viewer's attention. Using video formats that your audience is used to also breaks the monotony, accelerates your speech, and makes the editing more fun.

Regardless of the type of video you choose to make, the timing of its publication is very important. One criterion used by YouTube algorithms to promote your video is how fresh it is. Because of this, newer videos will get a small boost from the algorithms in the first few hours after publication.

Therefore, it is important to choose the timing of your uploads carefully. From this boost, YouTube analyzes whether your video has become popular quickly or slowly, and whether this popularity has been maintained. With this in mind, it is without question that both the video itself and the subject that it covers must be timed properly to become popular. Subjects that are trending at the time of publication are more likely to receive views.

Chapter Summary

This chapter covered the following topics:

- How algorithms on YouTube work and why it is important to keep audience retention rather than fishing for clicks.
- How to analyze data from your videos and the importance of assessing how people interact with them.
- Learning how to use the comments section to your advantage and increase the retention of each video.
- The importance of well-crafted descriptions and search-engine-optimized, flowing, informative text.
- The difference between long and short videos on YouTube and how to maximize your gain from each type, depending on your preferences.
- How publishing content at the right time and moment can be helpful to you and increase your reach.

Now that you understand a little more about how YouTube's algorithms work, you are ready to learn more about how to make money with your videos. That is our topic for the next chapter.

Chapter 5: How to Monetize a YouTube Channel

It's time to find out how to turn your hobby of creating videos into a job. But don't start thinking that you can live off your videos right away! YouTube does not monetize your videos for each upload, but rewards your efforts little by little, mainly through quick ads in between your content. Still, there are some conditions for you to receive this reward. The first step is to opt in to monetization for your content in "account settings." Then you can either join the YouTube Partner Program or list your videos on YouTube Premium.

YouTube Partner Program

Ok, but what exactly is the YouTube Partner Program? As the name implies, it is a partnership with the platform that gives you the possibility of compensation. But to become part of the program, you must first meet some conditions.

First, pay close attention to YouTube's community guidelines. These are the "ground rules of coexistence" for this social network. As you might imagine, things like spamming, spreading scams, impersonating others, creating false engagement, posting sensitive or violent content, selling weapons or other problematic goods, among other things, are practices that go against YouTube's policy and values and should be avoided. Before applying to take part in the YPP, make sure that all these rules are being followed.

Second, get acquainted with the monetization policies. This means that you can no longer scroll down that long Terms of Service page reading none of it. To comply with monetization policies, you must know what YouTube's Terms of Service says. You also need to get more information on their Copyright and Google AdSense program policies. Essentially, these state that all of your content must come from videos that you have produced yourself or that you have permission to use. You may not use other people's videos or videos that contain other people's work (such as music videos, scenes from TV shows, or the work of other professionals) without proper authorization.

In the YPP, you can incorporate ads into your videos, and each click that the ad receives generates engagement and rewards for you. However, if you use artificial means or special programs to produce clicks that do not come from genuine people or their genuine desire to learn more about the advertised product, then you are breaking the policies of the AdSense Program and will not receive proper monetization.

Likewise, you cannot bribe your viewers into clicking on your ads or offer compensation for your audience's engagement with your video ads. Even telling people to click on ads without offering compensation can have its price. Brands that advertise on YouTube expect to be truly rewarded for this, and if a person clicks on your product with no intention of buying it, you are being rewarded for something that adds nothing to that brand. You are monetizing on someone else's marketing without the brand getting anything in return. This is against the AdSense rules.

After you apply to join the YPP, YouTube checks a few details of your channel, such as the major theme, your most viewed and most recent videos, the highest watch time ratio, and video metadata (such as titles, thumbnails, and descriptions). Although YouTube's reviewers will not check your entire channel (after all, depending on the size it could take a long time), the verification process can be random and you don't know exactly what they will look at, so you need to comply with all community rules if you want to monetize your channel.

In addition, advertisers also may remove their ads from your videos if they feel that your content in any way hurts their brand image. You need to follow YouTube's policies not only to stay on the platform itself, but also to keep the trust of advertisers, and always remain respectful to everyone who has contact with your channel. You need to frequently check the YouTube policies pages to keep up to date, as any non-compliance with the rules can lead to demonetizing your videos or even banning you from the platform, ending your account altogether.

To monetize your videos, you also need to have some milestones completed, such as over 1,000 subscribers and 4,000 hours of public videos watched in the last 12 months. If you enable notifications in the "Monetization" tab of YouTube Studio, you will know when you are eligible to profit from your videos. When the time comes, you will also need to create an AdSense account to receive payments.

Unfortunately, the YPP is not available for all regions. Check if your country is on the list of eligible countries before you apply; you can find this list on YouTube itself.

YouTube Premium

YouTube Premium allows users to pay a small monthly fee to consume content without having to watch any ads anywhere on the platform. This means that it can also compensate you for each premium member who watches your videos.

You don't have to do anything yourself, as your videos are automatically visible to everyone in or out of premium membership. This is a secondary form of revenue that you receive alongside advertisements. The amount you get will depend on how many YouTube Premium members watch your content. You can view these numbers in YouTube Analytics. Members also have the option to download videos, which means they can watch your videos offline, allowing more watch time for your channel.

It may seem like a fairly simple way to get paid, but all guidelines must be taken into account for you to get any money from YouTube. So, if you do not comply with the rules or try to circumvent them, you also risk punishments that affect your premium revenue. You must read and follow YouTube's "virtual behavior" guidelines strictly if you want to receive any amount of money from them.

It is important to remember that YouTube's monetization policy is not perfect, and even though certain types of content are monetized today, nothing prevents YouTube from going back at some point and demonetizing your video or your channel.

And the opposite also happens. For example, YouTube does not monetize videos that have too many obscene words or refer to nudity or pornography, but this also means that it has sometimes demonetized completely educational and relevant videos on these subjects, such as sex or medical education videos.

The purpose of demonetizing videos is, besides ensuring content that does not harm the integrity of any user, to assure advertisers that their ads are not being associated with conspiracy theorists or people who might contribute to negatively affecting society. Again, it is very important to keep up to date with, and follow, YouTube's guidelines if you wish to be eligible for monetization.

Stream Live Videos

Making money on YouTube is not restricted to the ads that appear in your video. This is probably one of the least lucrative ways to earn money on YouTube — especially because of demonetization, which is often done in an automated way (although you may contest it). If you have established a solid subscriber base, with people who follow your work often and are loyal to your content, it is worth exploring the platform's live video streaming feature.

Streaming live videos is a way to connect with the people who consume your content, interact with them, understand why they follow you, and improve your work. But the best part is that live streaming videos on YouTube are also a way to earn money directly from your followers. This is because there is a tool that allows people to contribute to your channel by donating any amount they want — from 1 cent to thousands of dollars. Of course, it is unlikely that anyone will donate more than a hundred dollars on your live stream, but if you produce content that matches other people's reality and they can identify with you, then you can stand a chance of receiving a reasonable amount (or many small amounts from different people).

Here is the biggest reason to follow all the moral and bureaucratic guidelines of YouTube: when you are authentic, people will naturally come to you and support what you do. Learn to embrace this and your chances of growth will increase.

If you don't want to stream many live videos or if this is not exactly your thing, there is also the option of running a subscription service. This would be like receiving funding directly from your subscribers and fans of your work. People who respect, admire, and consider your content relevant may often want to financially support your channel's growth, and subscription services are perfect in these cases. One of the most popular platforms for this purpose is Patreon, which was even created by a YouTuber, Jack Conte.

You can also crowdfund your projects if you have the support of your viewers. For example, if you want to hire actors, buy better equipment, or record something that takes a little more work, and communicate this to your fans, crowdfunding can be a suitable alternative. For example, if you can offer your subscribers a little "sneak peek" of what's ahead to incite excitement, then they might be willing to donate money to bring the project to completion.

Launch Your Own Merchandise Line

If you want to make a living from YouTube, you need to find other ways of making money that go beyond the platform. A tactic widely used by many influencers is to launch their own merch. Creating your own products, such as t-shirts, mugs, stickers, or any other type of promotional item, can boost your channel financially and spread your content to other people outside the Internet. You can advertise it in your videos, through the description, or even by wearing or using the products while you film. Many YouTubers have found success with this method including MamaDoctorJones, Chloeandbeans, and others.

Advertising products on YouTube is so effective that it is the most lucrative method for both the creators themselves and for brands who want to spread the reach of their products. Just as TV commercials often use famous faces to represent their brands, the same goes for YouTubers. Influencers are the A-list celebrities of the Internet. After all, there are often hundreds, thousands, and even millions of people watching a single channel. But even if you are a D-list or even Z-list content producer, there is still hope for you. Partnering with brands is so lucrative that it deserves its own chapter. We will discuss this further in the next few pages.

Chapter Summary

This chapter was a brief introduction to some ways to make money through YouTube, including:

- Participating in the YouTube Partner Program.
- Profiting from YouTube Premium.
- Promoting live stream videos.
- Crowdfunding projects.
- Launching your own merchandise line.

Next, you will learn more about the most effective and profitable way to make money on YouTube: by establishing partnerships with brands.

Chapter 6: How to Partner with Brands

If you want to get rid of your day job and start making a substantial amount of money from YouTube, partnering with brands might be the key. Most millionaire YouTubers don't get their fortunes from random advertisements that appear in their videos, but from deals with small, medium, or large companies that aim to spread the word about their brands.

To give you an idea: according to Brendan Gahan, an expert in YouTube Marketing, it is estimated that an influencer receives USD 0.05 per view by including ads in their videos. But partnering with brands can give you a more lucrative deal depending on the particularities and reach of your specific niche. An average influencer usually charges about $20 per thousand subscribers. Considering that not all subscribers watch every video, we shouldn't expect to get 1000 views every time, so let's estimate your average views to be around 100 for your 1000 subscribers. If you make a partnership charging $20 for every thousand subscribers, that's already four times more than what you would get from video ads alone, for example.

This sounds wonderful, you must think, but where to start? Can you do it even though you are just a tiny little video creator lost in a sea of bigger influencers? Well, you certainly need to have a solid base of followers and faithful subscribers. You don't need to have a huge channel, but it certainly has to be big enough to take part in the YouTube Partner Program, for instance, and profit from advertising campaigns. Once you reach this level, do three important things: learn how to make a media kit, understand how to negotiate with brands, and know how to evaluate your work. In the sections below, we will describe each of these points so that you can maximize your potential with brands.

Create a Media Kit

The first step to closing partnerships is to create a media kit. Also known as a "press kit," this is not anything you need to buy, but a resume of sorts for influencers. Just like in regular CVs, it is a way of letting potential recruiters, who in this case are more like partners since there is no hierarchy, know more about you, your work, and especially your audience. There is, however, a difference compared to corporate CVs: the media kit must show your personality in such a way that it represents you in some aspect, exposes your values, and shares your worldview. If for job openings the candidates' profiles must be in line with the company's principles, in the world of YouTube and digital influencers, brands must be compatible with who you are and the type of content you post. It makes no sense to affiliate yourself with an adult brand if your channel is dedicated to children, for example.

Before you even start your media kit, pay attention to your target audience. For whom exactly are you producing content? If your work has a very specific audience (for instance, if you have a kid's channel), you will have a better chance of working with brands that sell products targeted at this same audience (say, companies that sell toys). Do you have a channel about women's health? Create a media kit targeting brands that benefit women. Is your channel about technology? Then try partnering with tech brands. Do you teach languages? Aim for companies that encourage bilingual education, and so on.

In YouTube's statistics, you can see exactly who is interested in your videos. Besides showing the age groups of your viewers, YouTube also shows the average view duration, attention percentage, and minutes watched for each of them. With this information in hand, you get a slightly better idea of who your audience is and what kinds of products they might be interested in. It's also worth paying attention to your audience's predominant gender and geographic location. Identify the highlights of all this data and add them to your media kit. This will increase the chances of partnering with the right brands.

Essentially, your media kit should contain four key points, these being: 1) information about your audience, 2) your statistics, 3) what kind of content you post, and 4) who you are. You should put all the major highlights of your channel in your media kit. Some people may find this very difficult or even impossible because their channels are not yet big, but regardless of the numbers, always ask yourself: what sets you apart? There's always *something* about you that will separate you from similar influencers.

For example, you may have a small channel, but your audience retention can still be high, with strong levels of interaction and videos that hold the attention of the few people who watch them. If people can watch your videos until the end, without closing the window/tab or skipping to the final minutes, then this is interesting to a brand because it suggests that your subscribers value your word.

Pay attention to the bigger picture of your channel. Review the statistics and write the highlights since they deserve to be emphasized in your media kit. Of course, the most essential information is the kind that adds the most value to you and your work. Even if your numbers are low, you may have more to offer, either in terms of retention or the fact that the way you produce your content is very relevant to a certain subject. For example, channels about medicine, history, architecture, or other professions may deserve recognition for their high quality, level of research, professionalism, and willingness to help others.

The way you choose to present your media kit to potential partners helps brands understand whether the product they are selling is a good fit for you. Knowing how to recognize your successes and which aspects of your channel can benefit certain brands can open up significant paths for your YouTube career. So, it doesn't matter if the numbers are low; brands can also invest in micro-influencers, as long as they show that they have something significant to add to their advertising campaigns.

If you are a lesser-known influencer, perhaps it's best not to focus on numbers, but on how you present yourself to brands. Don't be afraid to expose yourself and get out there. You will never know if you can get a partnership with a nice company if you don't step out of your comfort zone and venture into the modern world of business. If you feel apprehensive, ask yourself: what is the worst thing that could happen? Here's the answer, regardless of the circumstances: the brand could say that you don't have what they are looking for in a partnership at the moment, and you won't sign a contract with them. Would that be the end of the world? No, it wouldn't, because even if you don't have great potential at the time, brands can still save your media kit for the future and monitor you for other projects. If you grow in the meantime, they may contact you themselves, or you can always reach out again yourself.

This brings us to another important point: keep your media kit attractive and up-to-date. Maybe that means you need to invest in graphic design or pay someone to create a media kit for you, but keep in mind that you'll eventually earn something in return. In the world of social media, where appearances say a lot, a beautiful and well-presented media kit catches the attention of brands. Therefore, consider putting some time and effort into it; you are unlikely to regret it!

Learn to Detect a Good Partnership

To achieve the perfect partnership, it is essential to know yourself and what you offer to your audience and potential partner companies. Try to establish some partnership criteria. Do you prefer to partner only with brands that have something to do with the dominant theme of your channel? Or only brands that you already consume? Think about what is important to you when entering the world of advertising. This helps you filter out the companies you might not do well with.

Regardless of your partnership criteria, there is one thing that always works: being passionate about what you are saying. If you get the chance to talk about something that makes you as enthusiastic as you are with anything else on your channel, chances are that you will be a suitable partner for that brand. For this reason, whenever you think a brand will add to your public image and that the products they offer are things you would use in real life, it may be worth putting more effort into that company, because they are more likely to be a good fit. Besides, being committed conveys genuineness and transmits the idea that you believe in what you are selling (something your audience will recognize).

For example, let's say you love books of all kinds. You like them so much that your channel is almost solely about literature. If you get a partnership with a book publisher or even a small local bookstore, you might make the best ads about their products, and have a better chance of being able to increase their sales with the help of your audience.

Ultimately, it is always good to remember that even if you have a small channel with unimpressive numbers, you can still get interesting partnerships with brands that fit your content, so it's okay to create your partnership standards. If you are not sure what parameters to use to decide whether to make a partnership with a company, some agents can help you close deals. They could be really helpful if you don't know where to start, or if you feel you're walking into a minefield! Besides, these are experts, and they will help you negotiate more realistic offers. But before researching potential agents, it may be a good idea to create a table or a list of acceptable and unacceptable conditions for working with someone. With this, you can decide what you find tolerable or intolerable (both in terms of payment and content). This helps you create your parameters without having to compare yourself with other influencers.

Still, don't despair if you can't afford an agent, or if at first you only come into contact with brands that are not so aligned with your way of doing things. For many, the power of choice comes with time. Small influencers often feel forced to pursue partnerships with brands they don't believe in because they think they need them in order to grow. When analyzing each offer, it's okay to consider your particular situation, but the most important thing is that you are comfortable with any agreement you may close, and with whatever you are proposing to do — including all the risks that may be involved. Keep in mind that believing in the brand and knowing that it fits your content and your audience makes a big difference, so keep looking for that whenever you have the chance!

Negotiate with Brands

Once you have your media kit ready, and assuming your channel is attractive enough, you will start receiving proposals from companies. Make sure you let people know the best way to contact you; that means including your corporate email address in your social networks. This way, even companies that haven't seen your media kit, but identify with your content, will be able to contact you for a potential deal. Remember not to get carried away with the excitement of finally getting a partnership! Accepting everything that comes your way may not be so good for you or your public image. If you accept an offer, you must always have all negotiations documented in writing, even if it is via email. Every kind of advertising negotiation must have a contract. If it doesn't, the company will be free to use you as they see fit and do things for their benefit. Remember; you are a professional, not a puppet!

For example, some brands like to send gifts to influencers for promoting their products. This kind of exchange also needs to be agreed upon. Some brands refuse to negotiate and still send you gifts; in this case, it is best not to promote the gift to your followers, as you would be doing something for free and without prior agreement. If you think it is worthwhile to advertise the product for future partnerships, make sure you establish a deal or an understanding with the company. This shows professionalism and self-worth.

Another thing that many influencers forget, but which is of extreme importance, is to signal your ads. If you receive proposals that suggest you don't need to show that the content is associated with a brand, it is best to turn them down. You should always tell subscribers when a video is directly associated with a company or when it is a paid partnership. This shows transparency with your audience, and it doesn't take much effort either — all you need to do is include a hashtag in your title ("#ad").

Knowing your limits and understanding which brands you want to engage with can make all the difference. Often a financial offer can tempt, but it can also damage your image, cross a line with your values, or harm your work. Once you become an influencer, you also become a brand. Try to understand what your brand is, what exactly it means, and what it would represent to a company that wants to be associated with it. If in doubt, try thinking of your audience: if you were one of your subscribers, would you want to consume this product? People don't subscribe to channels to see advertisements. If you partner with a lot of brands and most of your videos are ads, your channel could turn into something superficial and consumer-driven, which could cause your number of followers to drop. It is important to know exactly what you want to achieve with this kind of partnership and not sell yourself for any price. Sticking to your essence is worth more than any money — it's the reason people follow you and keep consuming what you produce. If that ends, so does the hype, and therefore the profit.

When closing a contract, pay close attention to each clause. It is not unusual for a company to include a paragraph that forbids the influencer to speak ill of the brand, even after the contract is terminated. If this is the case, it is worth making an amendment and including an additional clause that ensures that, in the event the company or the brand associated with it contradicts your values and principles, you may take a stand against it. By doing so, you will end the contract with no penalties, and if in the future this brand is associated with some sort of moral, corrupt, or unacceptable scandal in your eyes, you will not have your freedom of speech affected, because the contract allows you to express your thoughts freely.

When negotiating with any brand, remember to trust yourself and study each offer carefully. With time you will be more acquainted with what works for you, and you will become more proficient at negotiating contracts.

Read and Discuss Briefings

Already have your media kit and negotiated a nice proposal? Then you are likely to receive a briefing from the company in the next step. This is nothing more than some basic guidelines or instructions on how you should conduct the advertising of the product. The brand may ask you to behave in a certain way or to include certain "keywords" or specific phrases in your video.

Therefore, before you even close a deal with a particular company, ask about the influencer brief and what the company expects you to talk about when promoting the brand. You should always know beforehand how much freedom you have over the content and if you must phrase things exactly as they instruct, etc. It is important that when you have to word things specifically, that everything is very consistent with what you believe.

Once you have the marketing brief ready, read it carefully and understand what the most important aspects of the brand are. These are the things that you cannot change because the company needs them in their ad. If you have agreed on some creative freedom when doing the advertisement, you can also create something that fits the mold of your channel. For example, let's say you have a channel about music and recently established a partnership to promote a product. The partner company asks you to say a few phrases, but gives you complete freedom to express them however you like. You choose to follow the marketing brief by creating a melody and a song for your advertisement. This is the value you have to add to the product: putting your brand, your personality, and your digital identity in combination with the company's requirements.

No matter what your final decision is, keep in mind that honesty is the key to success; if you are advertising a product but do not consume it, do not say that you use it every day. Instead, you can say that you were invited to test the product. This prevents you from lying to your audience and strengthens the level of trust between you and your viewers. Make sure that the influencer brief does not make it impossible for you to remain sincere with your subscribers. They will thank you for it!

Estimate Your Worth

One of the biggest challenges for any self-employed person is to assess the value of their job. When you work on YouTube, chances are that you are taking the role of every professional that would be involved in video creation. So don't be afraid to do the math! When closing an advertising partnership, consider the time and services it will take you to promote their brand.

Think about each role you will have to assume to follow the briefing shared by your partner company. Include in this list the number of hours it should take for you to prepare the equipment, the lighting, and the general setting of your video, as well as the shooting time itself. How much time will you take to create a script, edit the video, and create a thumbnail? Your entire creative process needs to be priced and explained to the brand.

Account for the minimum amount that each of these professionals (that's you) should earn for their work, plus the value of advertising your content and the product. Once you figure out the minimum hourly wage of each of those jobs, that will be the minimum amount that you should charge the partner company. When you learn to negotiate these terms with the brand, you show potential partners that you also value yourself and your work as a creator.

Advertising should align the best interests of the three pillars of digital influencing: the brand, the creator, and the audience. The result should be consistent with what is expected by all three. Explain your goals and what your audience expects from you to your partner company so that you can reach a final understanding of what works best for everyone involved. Nobody can afford to lose, and no one understands your audience as well as you do. When you do this, your ads will reach the audience in a much more authentic and transparent way.

Chapter Summary

This last chapter covered in a little more depth the most important aspects to keep in mind when partnering with brands. Here, you found out more about:

- The importance of developing, enhancing, and updating a media kit to attract potential partner brands that may have similar values to yours.
- Important aspects to consider when trying to spot a good partnership.
- How to negotiate with brands that spark your interest, the importance of establishing written agreements and of including clauses that protect you as a creator and guard your freedom of expression.
- The specifics of the marketing briefing process and the importance of a close reading of its contents, as well as the possibilities of negotiating changes so that whatever is promoted is in line with your interests.
- How to estimate the value of your work when interested brands request a quotation to close the partnership.

Final Words

If you have read this book from start to finish, then hopefully you have found answers to your key questions about the world of YouTube content creation. But don't worry if you skipped some parts and just focused on what matters to you now, the information here is for you whenever you need it! I designed this book just for that purpose, in such a way that you can visit and revisit it every time an additional question arises.

If there is one thing that every step of the YouTube process has shown in this book, it is that monetizing from YouTube is important if you want to establish yourself on the platform, but this monetization cannot be sustained without your dedication, passion, and honesty with your audience. Hopefully, these tips have added something to your creative process, and you can continue to create authentic videos that are true to your essence.

Now that you know these little secrets, what are you waiting for? Time for you to get to work, and we'll see you on YouTube!

References

Cooper, Page. How Does the YouTube Algorithm Work? A Guide to Getting More Views. Hootsuit, accessed May 3, 2021. Available at: https://blog.hootsuite.com/how-the-youtube-algorithm-works/

Fairweather, Lauren, Shepherd, Will. What is Branded Content? In: YouTube Creators. Accessed April 29, 2021. Available at: https://www.youtube.com/watch?v=6ysdXcdKy-4

Ho, Leslie. How to Write the Perfect Influencer Marketing Brief. Inzpire.me, accessed May 7, 2021. Available at: How to Write the Perfect Influencer Marketing Brief - inzpire.me Blog

Hogoboom, Emily. The 7 Things Big Brands Want to See in your Influencer Media Kit. The Shelf, accessed May 14, 2021. Available at: https://www.theshelf.com/influencer-resources/influencer-media-kit

Jaffari, Amir. How the YouTube Algorithm Works (Or Why Your Videos Aren't Getting Views). Shopify Blog: Accessed May 16, 2021. Available at: https://www.shopify.com/blog/youtube-algorithm

Kumar, Braveen. How to Make Money on YouTube (Without a Million Subscribers). Shopify Blog, accessed May 15,

2021. Available at: https://www.shopify.com/blog/198134793-how-to-make-money-on-youtube

Odjick, Desirae. Like, Comment, and Thrive: How to Start a Successful YouTube Channel for Your Business. Shopify Blog, accessed May 16, 2021. Available at: Quickstart Guide: Start a YouTube Channel for Your Business (2021) (shopify.com)

YouTube Video Equipment List: What You Need to Start Recording Videos. Adorama, accessed May 10, 2021. Available at: https://www.adorama.com/alc/youtube-video-equipment-list-what-you-need-to-start-recording-videos/

www.ingramcontent.com/pod-product-compliance
Lightning Source LLC
Chambersburg PA
CBHW071513210326
41597CB00018B/2742